Connected Mathematics

How Likely Is It?

Probability

Student Edition

Glenda Lappan
James T. Fey
William M. Fitzgerald
Susan N. Friel
Elizabeth Difanis Phillips

Developed at Michigan State University

DALE SEYMOUR PUBLICATIONS

The Connected Mathematics Project was developed at Michigan State University with the support of National Science Foundation Grant No. MDR 9150217.

This project was supported, in part, by the
National Science Foundation
Opinions expressed are those of the authors and not necessarily those of the Foundation

The Michigan State University authors and administration have agreed that all MSU royalties arising from this publication will be devoted to purposes supported by the Department of Mathematics and the MSU Mathematics Education Enrichment Fund.

This book is published by Dale Seymour Publications®, an imprint of the Alternative Publishing Group of Addison-Wesley Publishing Company.

Managing Editor: Catherine Anderson
Project Editor: Stacey Miceli
Book Editor: Mali Apple
Production/Manufacturing Director: Janet Yearian
Production/Manufacturing Coordinator: Claire Flaherty
Design Manager: John F. Kelly
Photo Editor: Roberta Spieckerman
Design: PCI, San Antonio, TX
Composition: London Road Design, Palo Alto, CA
Illustrations: Pauline Phung, Margaret Copeland, Ray Godfrey
Cover: Ray Godfrey

Many of the designations used by manufacturers and sellers to distinguish their products are claimed as trademarks. Where those designations appear in this book, and Dale Seymour Publications® was aware of a trademark claim, the designations have been printed in initial caps or all caps.

Photo Acknowledgements: 11 © Harold M. Lambert/Superstock, Inc.; 17 © John Moore/The Image Works; 27 © Hazel Hankin/Stock, Boston; 39 © Michael Dwyer/Stock, Boston; 53 © Ed Fowler/UPI/Bettmann Newsphotos; 55 © Michael Tamborrino/FPG International; 57 © Dennis E. Cox/Tony Stone Images, Inc.; 58 © Michael Weisbrot/Stock, Boston

Copyright © 1996 by Michigan State University, Glenda Lappan, James T. Fey, William M. Fitzgerald, Susan N. Friel, and Elizabeth D. Phillips. All rights reserved. No part of this publication may be reproduced, stored in a retrieval system, or transmitted, in any form or by any means, electronic, mechanical, photocopying, recording, or otherwise, without prior written permission of the authors. Printed in the United States of America.

Monopoly is a trademark of Parker Brothers.

DALE SEYMOUR PUBLICATIONS®
P.O. BOX 10888
PALO ALTO, CA 94303

Order number 21451
ISBN 1-57232-156-3

1 2 3 4 5 6 7 8 9 10-BA-99 98 97 96 95

The Connected Mathematics Project Staff

Project Directors

James T. Fey
University of Maryland

William M. Fitzgerald
Michigan State University

Susan N. Friel
University of North Carolina at Chapel Hill

Glenda Lappan
Michigan State University

Elizabeth Difanis Phillips
Michigan State University

Project Manager

Kathy Burgis
Michigan State University

Technical Coordinator

Judith Martus Miller
Michigan State University

Curriculum Development Consultants

David Ben-Chaim
Weizmann Institute

Alex Friedlander
Weizmann Institute

Eleanor Geiger
University of Maryland

Jane Mitchell
University of North Carolina at Chapel Hill

Anthony D. Rickard
Alma College

Collaborating Teachers/Writers

Mary K. Bouck
Portland, Michigan

Jacqueline Stewart
Okemos, Michigan

Graduate Assistants

Scott J. Baldridge
Michigan State University

Angie S. Eshelman
Michigan State University

M. Faaiz Gierdien
Michigan State University

Jane M. Keiser
Indiana University

Angela S. Krebs
Michigan State University

James M. Larson
Michigan State University

Ronald Preston
Indiana University

Tat Ming Sze
Michigan State University

Sarah Theule-Lubienski
Michigan State University

Jeffrey J. Wanko
Michigan State University

Evaluation Team

Diane V. Lambdin
Indiana University

Sandra K. Wilcox
Michigan State University

Judith S. Zawojewski
National-Louis University

Teacher/Assessment Team

Kathy Booth
Waverly, Michigan

Anita Clark
Marshall, Michigan

Theodore Gardella
Bloomfield Hills, Michigan

Yvonne Grant
Portland, Michigan

Linda R. Lobue
Vista, California

Suzanne McGrath
Chula Vista, California

Nancy McIntyre
Troy, Michigan

Linda Walker
Tallahassee, Florida

Software Developer

Richard Burgis
East Lansing, Michigan

Development Center Directors

Nicholas Branca
San Diego State University

Dianne Briars
Pittsburgh Public Schools

Frances R. Curcio
New York University

Perry Lanier
Michigan State University

J. Michael Shaughnessy
Portland State University

Charles Vonder Embse
Central Michigan University

Special thanks to the students and teachers at these pilot schools!

Baker Demonstration School
Evanston, Illinois

Bertha Vos Elementary School
Traverse City, Michigan

Blair Elementary School
Traverse City, Michigan

Bloomfield Hills Middle School
Bloomfield Hills, Michigan

Brownell Elementary School
Flint, Michigan

Catlin Gabel School
Portland, Oregon

Cherry Knoll Elementary School
Traverse City, Michigan

Cobb Middle School
Tallahassee, Florida

Courtade Elementary School
Traverse City, Michigan

Duke School for Children
Durham, North Carolina

DeVeaux Junior High School
Toledo, Ohio

East Junior High School
Traverse City, Michigan

Eastern Elementary School
Traverse City, Michigan

Eastlake Elementary School
Chula Vista, California

Eastwood Elementary School
Sturgis, Michigan

Elizabeth City Middle School
Elizabeth City, North Carolina

Franklinton Elementary School
Franklinton, North Carolina

Frick International Studies Academy
Pittsburgh, Pennsylvania

Gundry Elementary School
Flint, Michigan

Hawkins Elementary School
Toledo, Ohio

Hilltop Middle School
Chula Vista, California

Holmes Middle School
Flint, Michigan

Interlochen Elementary School
Traverse City, Michigan

Los Altos Elementary School
San Diego, California

Louis Armstrong Middle School
East Elmhurst, New York

McTigue Junior High School
Toledo, Ohio

National City Middle School
National City, California

Norris Elementary School
Traverse City, Michigan

Northeast Middle School
Minneapolis, Minnesota

Oak Park Elementary School
Traverse City, Michigan

Old Mission Elementary School
Traverse City, Michigan

Old Orchard Elementary School
Toledo, Ohio

Portland Middle School
Portland, Michigan

Reizenstein Middle School
Pittsburgh, Pennsylvania

Sabin Elementary School
Traverse City, Michigan

Shepherd Middle School
Shepherd, Michigan

Sturgis Middle School
Sturgis, Michigan

Terrell Lane Middle School
Louisburg, North Carolina

Tierra del Sol Middle School
Lakeside, California

Traverse Heights Elementary School
Traverse City, Michigan

University Preparatory Academy
Seattle, Washington

Washington Middle School
Vista, California

Waverly East Intermediate School
Lansing, Michigan

Waverly Middle School
Lansing, Michigan

West Junior High School
Traverse City, Michigan

Willow Hill Elementary School
Traverse City, Michigan

Contents

Mathematical Highlights   4

Investigation 1: A First Look at Chance   5
    1.1 Flipping for Breakfast   5
    1.2 Analyzing Events   7
    Applications—Connections—Extensions   9
    Mathematical Reflections   13

Investigation 2: More Experiments with Chance   14
    2.1 Tossing Marshmallows   14
    2.2 Pondering Possible and Probable   16
    Applications—Connections—Extensions   17
    Mathematical Reflections   21

Investigation 3: Using Spinners to Predict Chances   22
    3.1 Bargaining for a Better Bedtime   22
    Applications—Connections—Extensions   24
    Mathematical Reflections   28

Investigation 4: Theoretical Probabilities   29
    4.1 Predicting to Win   29
    4.2 Drawing More Blocks   32
    4.3 Winning the Bonus Prize   33
    Applications—Connections—Extensions   35
    Mathematical Reflections   41

Investigation 5: Analyzing Games of Chance   42
    5.1 Playing Roller Derby   42
    Applications—Connections—Extensions   44
    Mathematical Reflections   48

Investigation 6: More About Games of Chance   49
    6.1 Scratching Spots   49
    Applications—Connections—Extensions   51
    Mathematical Reflections   56

Investigation 7: Probability and Genetics   57
    7.1 Curling Your Tongue   57
    7.2 Tracing Traits   58
    Applications—Connections—Extensions   61
    Mathematical Reflections   64

How Likely Is It?

Imagine that you are one of four contestants on a game show. The host is holding a bucket containing red, yellow, and blue blocks. You cannot see the blocks. Each contestant guesses a color and then draws a block from the bucket. A contestant who correctly predicts the color of the block wins $500. After each draw, the block is returned to the bucket. What are your chances of winning the game? Is there an advantage to choosing first? Is there an advantage to choosing last?

Suppose you have a scratch-off game card with five spots. Each spot covers the name of a prize. The names under two of the spots match. You are allowed to scratch off two spots. If the names under the spots match, you win the prize. How likely is it that you will win?

Some people can curl their tongues into a U shape; other people can't. What are the chances that a person can curl her or his tongue?

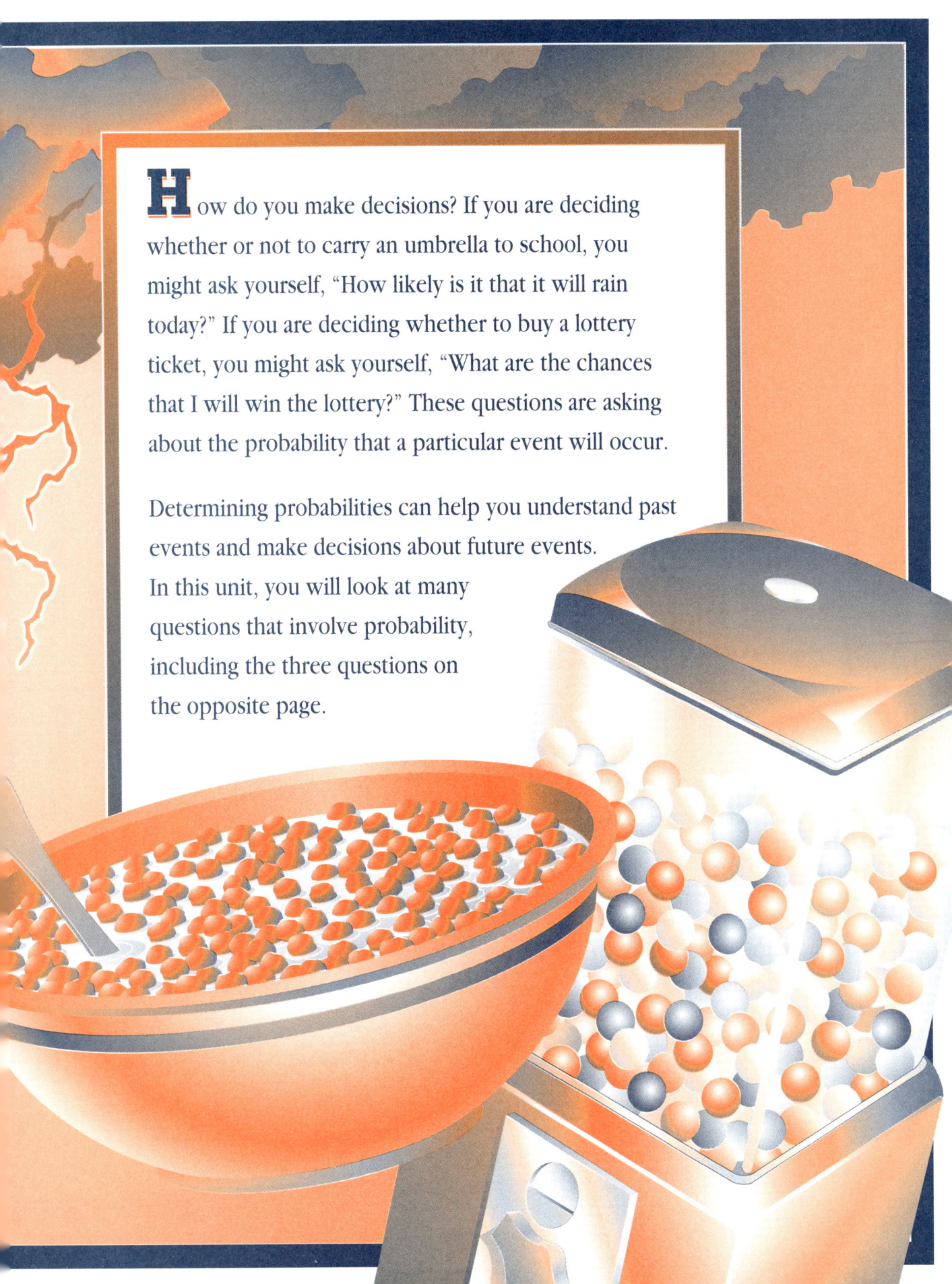

How do you make decisions? If you are deciding whether or not to carry an umbrella to school, you might ask yourself, "How likely is it that it will rain today?" If you are deciding whether to buy a lottery ticket, you might ask yourself, "What are the chances that I will win the lottery?" These questions are asking about the probability that a particular event will occur.

Determining probabilities can help you understand past events and make decisions about future events. In this unit, you will look at many questions that involve probability, including the three questions on the opposite page.

Mathematical Highlights

In *How Likely Is It?* you will learn about probability.

- Analyzing the results of experiments with coins, spinners, blocks, and marshmallows allows you to make predictions about what will happen over the long run.

- Considering the possible results of several actions helps you understand the difference between equally likely and unequally likely events.

- Experiments with games of chance illustrate the difference between possible and probable events.

- Analyzing all the possible resulting events of a game shows you how to use models to determine probabilities mathematically rather than by experimenting.

- Making an organized chart of all the possible outcomes of rolling two number cubes helps you develop strategies for winning a game.

- Simulating a scratch-off card game lets you determine the chances of winning such a contest.

- Using some information about genetic traits lets you see how probabilities can be used to predict whether a person will have a specific trait, such as tongue-curling ability.

INVESTIGATION 1

A First Look at Chance

One way to make a decision about something is to do an experiment to see what is likely to happen. In this investigation, you will experiment with flipping a coin.

1.1 Flipping for Breakfast

Kalvin, an eighth grader, always has cereal for breakfast. He likes Cocoablast cereal so much that he wants to eat it every morning. Kalvin's mother wants him to eat Health Nut Flakes at least some mornings because it is more nutritious than Cocoablast.

Kalvin and his mother have come up with a fun way to determine which cereal Kalvin will have for breakfast. Each morning, Kalvin flips a coin. If the coin comes up heads, he will have Cocoablast. If he flips a tail, he will have Health Nut Flakes.

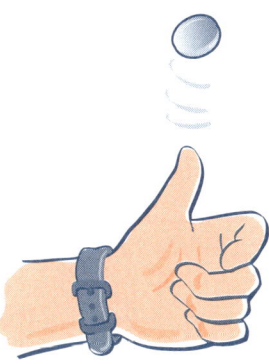

Problem 1.1

How many days in June do you think Kalvin will eat Cocoablast?

Explore this question by flipping a coin 30 times to determine Kalvin's cereal for each morning in June. Use Labsheet 1.1 to help you collect your data.

June						
	1	2	3	4	5	6
7	8	9	10	11	12	13
14	15	16	17	18	19	20
21	22	23	24	25	26	27
28	29	30				

For each day, record the result of the flip (H or T) and the percent of heads so far. Use the data to make a coordinate graph with the days from 1 to 30 on the *x*-axis and the percent of heads so far on the *y*-axis.

■ **Problem 1.1 Follow-Up**

Work with your teacher to combine the results from all the groups.

1. **a.** What fraction of the entire class's flips were heads?
   b. As you added more and more data, did the fraction of heads get closer to or further from $\frac{1}{2}$?
2. **a.** Based on what you found for June, how many times would you expect Kalvin to eat Cocoablast cereal in July?
   b. How many times would you expect Kalvin to eat Cocoablast cereal in a year?
3. Kalvin's mother told him that the chances of getting a head when you flip a coin are $\frac{1}{2}$. Does this mean that every time you flip a coin twice you will get one head and one tail? Explain your reasoning.

1.2 Analyzing Events

Kalvin found a penny near a railroad track. It looked flattened and a bit bent, so Kalvin assumed it must have been run over by a train. He decided to use this unusual penny for determining his breakfast.

Kalvin's mother became suspicious of the penny at the end of June because Kalvin had eaten Health Nut Flakes only seven times. She explained why she was suspicious. "With a fair coin, heads and tails are **equally likely** results. This means that you have the same chance of getting a head as a tail. I just don't think your coin is fair!"

Think about this!

Do you think heads and tails are equally likely with Kalvin's penny? How could Kalvin find out whether his coin is fair?

Kalvin was not quite sure what his mother meant by *equally likely*, so she made up an example to help explain it.

"Suppose everyone in our family wrote his or her name on a card and put the card in a hat. If you mixed up the cards and pulled one out, each name would have an equally likely chance of being picked. But suppose I put my name in the hat ten times. Then when you picked one card out of the hat, our names wouldn't all have an equal chance of being picked—my name would have a greater chance of being chosen than everyone else's name."

Investigation 1: A First Look at Chance

Problem 1.2

In A–H, decide whether the possible resulting events of each action are equally likely, and briefly explain your answer.

Action	Possible resulting events
A. You toss a soda can.	The can lands on its side, the can lands upside down, or the can lands right side up.
B. You roll a number cube.	1, 2, 3, 4, 5, or 6
C. You check the weather in Alaska on a December day.	It snows, it rains, or it does not rain or snow.
D. The Pittsburgh Steelers play a football game.	The Steelers win, the Steelers lose, or the Steelers tie.
E. A baby is born.	The baby is a boy or the baby is a girl.
F. A baby is born.	The baby is right-handed or the baby is left-handed.
G. You guess on a true/false question.	The answer is right or the answer is wrong.
H. You shoot a free throw.	You make the basket or you miss.

■ **Problem 1.2 Follow-Up**

1. Describe three other situations in which the possible resulting events are equally likely.

2. Describe three other situations in which the possible resulting events are not equally likely.

Applications • Connections • Extensions

As you work on these ACE questions, use your calculator whenever you need it.

Applications

1. a. Sarah flipped a coin 50 times, and heads turned up 28 times. What fraction of the 50 flips of the coin turned up heads?

   b. If the coin is fair, and Sarah flips it 500 times, how many times should she expect it to come up heads?

2. Suppose Kalvin flipped a coin to determine his breakfast cereal every day starting on his twelfth birthday and continuing until his eighteenth birthday. How many times would you expect him to eat Cocoablast cereal?

3. Kalvin flipped a coin five days in a row and got tails every time. He told his mother there must be something wrong with the coin he was using. Do you think there is something wrong with the coin? How could Kalvin find out?

4. Len flipped a coin three times and got a head each time. What are the chances he will get a tail on his next toss? Explain your reasoning.

5. Is it possible to flip a coin 20 times and have it turn up heads 20 times? Is this likely to happen? Explain your reasoning.

Investigation 1: A First Look at Chance

Connections

In 6–10, decide whether the resulting events are equally likely, and briefly explain your answer.

Action	Possible resulting events
6. Your phone rings at 9:00 P.M.	The caller is your best friend, the caller is a relative, or the caller is someone else.
7. You check the temperature in your area tomorrow.	The temperature is over 30°F or the temperature is under 30° F.
8. You spin this spinner.	The spinner lands on stripes, the spinner lands on hearts, or the spinner lands on dots.
9. Your teacher arrives at school in the morning.	Your teacher arrives on time or your teacher is late.
10. You find out how many car accidents occurred in your city or town yesterday.	There were fewer than five accidents, there were exactly five accidents, or there were more than five accidents.

In 11–14, use this graph, which shows the average number of tornadoes per year in several states:

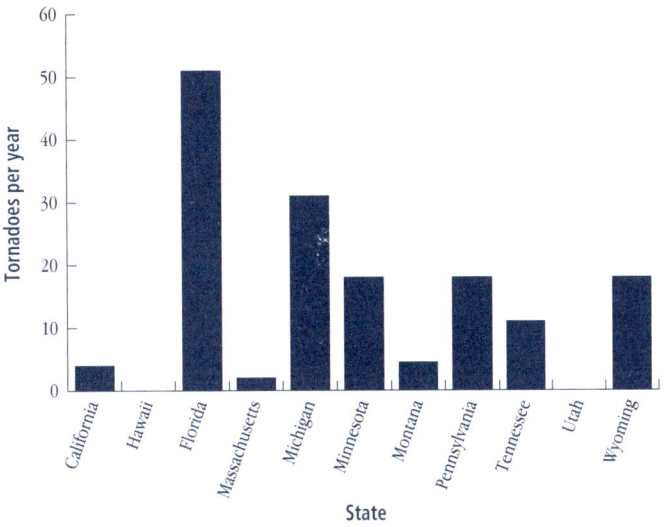

11. Is it equally likely for a tornado to hit somewhere in California as for a tornado to hit somewhere in Florida?

12. Is it equally likely for a tornado to hit somewhere in Minnesota as for a tornado to hit somewhere in Pennsylvania?

13. Is it equally likely for a tornado to hit somewhere in Massachusetts as for a tornado to hit somewhere in California?

14. Based on these data, is a person living in Montana in more danger of being hit by a tornado than a person living in Massachusetts? Explain your reasoning.

Investigation 1: A First Look at Chance

Extensions

15. Monday was the first day Kalvin flipped a coin to determine his cereal. During his first five days of flipping, he only had Cocoablast twice. One way that Kalvin could have done this was to have flipped heads on Monday and Tuesday and tails on Wednesday, Thursday, and Friday. We can write this as:

Monday	Tuesday	Wednesday	Thursday	Friday
H	H	T	T	T

Find every other way Kalvin could have flipped the coin during the week and had Cocoablast cereal twice. Explain how you know that you have found every way.

Mathematical Reflections

In this investigation, you experimented with coins to determine the fraction of heads and tails that occurred when you tossed a coin 30 times and when you combined the tosses from all the students in your class. You also investigated other situations to evaluate whether the possible resulting events were equally likely. These questions will help you summarize what you have learned:

1. What does it mean to say that the chances of getting a head when a coin is tossed are $\frac{1}{2}$?

2. If you experiment by tossing a coin and tallying the results, are 30 tosses as good as 500 tosses to predict the chances of a coin landing tails up? Explain why or why not.

3. a. What does it mean for the results of some action to be equally likely?

   b. Give an example of an action in which the possible resulting events are equally likely.

   c. Give an example of an action in which the possible resulting events are not equally likely.

4. If you toss a fair coin, is it *possible* to get 25 heads in a row? Is this *likely* to happen?

   Think about your answers to these questions, discuss your ideas with other students and your teacher, and then write a summary of your findings in your journal.

INVESTIGATION 2

More Experiments with Chance

Kalvin loves Cocoablast cereal so much that he wants to find something else to flip that will give him a better chance of eating it each morning.

2.1 Tossing Marshmallows

Kalvin looked through the kitchen cupboard and found a bag of large marshmallows and a bag of small marshmallows. He thought that a marshmallow might be a good thing to flip, and wondered which size would be better. Since Kalvin wants to eat Cocoablast most of the time, he needs to find a marshmallow that lands in one position—either on its side or on one of its flat ends—most of the time. Once he decides which type of marshmallow is better, he will ask his mother if he may use the marshmallow instead of a coin for deciding his cereal each morning.

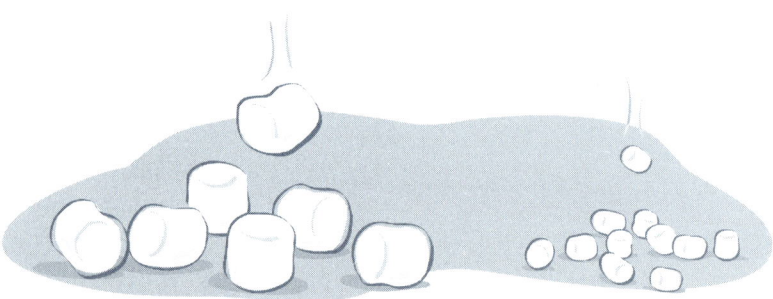

Did you know?

Originally, marshmallows were made from the root of the marsh mallow, a pink-flowered European perennial herb. Today, most marshmallows are made from corn syrup, sugar, albumen, and gelatin.

Problem 2.1

Experiment with large and small marshmallows to help you answer these questions:

A. Which size marshmallow should Kalvin use to determine which cereal he will eat? Explain your answer.

B. Which of the marshmallow's landing positions—end or side—should Kalvin use to represent Cocoablast? Explain your answer.

To conduct your experiment, toss each size of marshmallow 50 times. Keep track of your data carefully. Here is an example of how you might want to organize your data:

	Lands on an end	Lands on side
Large marshmallow	‖‖‖ ‖	‖‖‖
Small marshmallow		

Use the results of your experiment to help you answer questions A and B.

■ **Problem 2.1 Follow-Up**

Work with your teacher to combine results from all the groups.

1. **a.** For what fraction of your 50 tosses did the large marshmallow land on one of its ends? On its side?
   b. For what fraction of the class's tosses did the large marshmallow land on one of its ends? On its side?
   c. If you toss a large marshmallow once each day for a year, how many times would you expect it to land on its side?

2. **a.** For what fraction of your 50 tosses did the small marshmallow land on one of its ends? On its side?
   b. For what fraction of the class's tosses did the small marshmallow land on one of its ends? On its side?
   c. If you toss a small marshmallow once each day for a year, how many times would you expect it to land on its side?

3. Suppose Kalvin uses the marshmallow you chose—large or small—to decide his cereal each morning. He tosses the marshmallow twice, and it lands on an end once and on its side once. He says, "This marshmallow isn't any better than the penny—it lands on an end 50% of the time!" How would you convince Kalvin that the marshmallow is better for him to use than a penny?

Investigation 2: More Experiments with Chance

2.2 Pondering Possible and Probable

Jon and Tat Ming are playing a coin-tossing game. To play the game, they take turns tossing three coins. If all three of the coins match, Jon scores a point. If only two of the coins match, Tat Ming scores a point. The first player to get 5 points wins. Both players have won the game several times, but Tat Ming seems to be winning more often. Jon says that he thinks the game is unfair. Tat Ming claims that the game is fair because both of them have a chance to win.

What do you think? Is the game fair as long as it is possible for each player to win?

Problem 2.2

Conduct an experiment to help you answer these questions:

A. Is it possible for Jon to win the game? Is it possible for Tat Ming to win the game? Explain your reasoning.

B. Who is more likely to win? Why?

C. Is this a fair game of chance? Explain.

To conduct your experiment, toss three coins 30 times. Keep track of the number of times three coins match and the number of times only two coins match. Be sure to organize your data and give reasons for your conclusions.

■ Problem 2.2 Follow-Up

1. If you tossed the coins 30 more times, how many times would you expect the three coins to match?
2. Toss the coins 30 more times. Compare this set of results to your first set of results. Did the three coins match about the same number of times in each experiment?

Applications • Connections • Extensions

As you work on these ACE questions, use your calculator whenever you need it.

Applications

1. When you toss a marshmallow, are the chances that it will land on an end the same as the chances that it will land on its side? That is, are the two events equally likely? Explain your reasoning.

2. If Kalvin uses the size marshmallow that you chose in Problem 2.1, how many times a month would you expect him to eat Cocoablast? How many times a year? Explain your reasoning.

3. Dawn tossed a pawn from her chess set 5 times. It landed on its base 4 times and on its side only once. Dawn decided that the pawn lands on its base more often than on its side.

   Andre tossed the same pawn 100 times. It landed on its base 28 times and on its side 72 times. Andre decided the pawn lands on its side more often than its base.

   Based on Andre and Dawn's data, if you toss the pawn one more time, do you think it would be more likely to land on its base or its side? Why?

Connections

4. Meteorologists make many claims about the chances of rain, sun, and snow occurring. Waldo, the meteorologist from WARM radio, claims he is the best weather predictor in Sunspot, South Carolina. On the day before Sunspot High's graduation ceremony, Waldo said: "There is only a 10% chance of rain tomorrow!"

a. Ask at least two adults what they think Waldo's statement means, and write down their explanations.

b. Explain what you think Waldo's statement means.

c. If it rains on the graduation ceremony, was Waldo wrong? Why or why not?

You can use a fraction or a percent to indicate the chances that a particular event will occur. The larger the fraction or percent, the greater the chances that the event will happen. If an event is impossible, the chances that it will occur are 0, or 0%. If an event is sure to happen, the chances that it will occur are 1, or 100%.

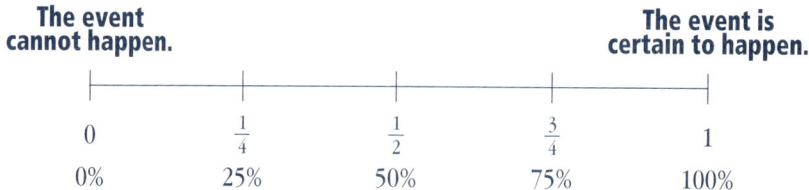

In 5–12, assign a number from 0 to 1 to indicate the chances that the event will occur, and explain your reasoning. For example, if the event is, "You will watch television tonight," your answer might be this:

I watch some television every night unless I have too much homework. So far today I do not have much homework. Therefore, I am about 95% sure that I will watch television tonight.

5. You will be absent from school at least one day during this school year.

6. You will have pizza for lunch one day this week.

7. It will snow on July 4 this year in Mexico.

8. You will get all the problems on your next math test correct.

9. The next baby born in your local hospital will be a girl.

10. The sun will set tonight.

11. You will win a coin-tossing game by tossing four coins, all of which must land heads.

12. You will toss a coin and get 100 tails in a row.

13. Make up two of your own events, and then estimate the chances that each event will happen.

In 14–16, use the chart below, which shows the percent of people who have been fired from a job for various reasons.

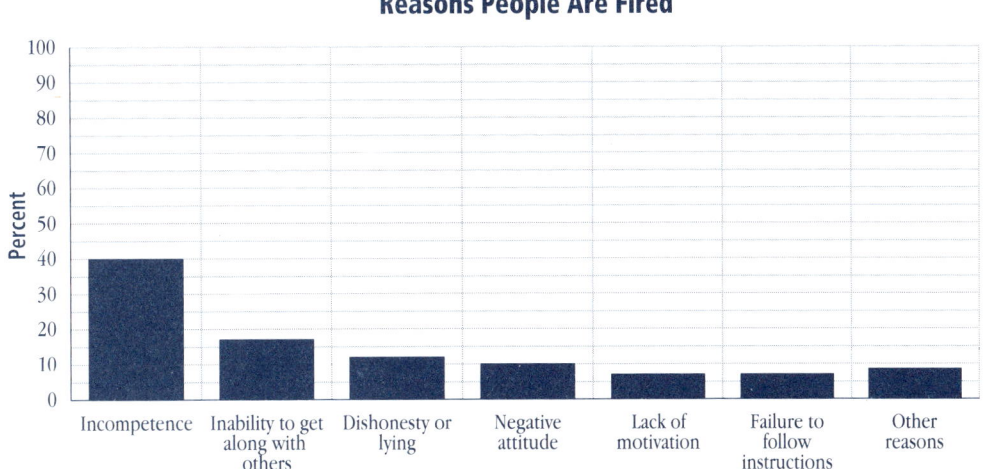

Source: Michael D. Shook and Robert L. Shook, *The Book of Odds* (New York: Penguin books, 1991), p. 53.

14. If this chart represents 5000 people, about how many of these people were fired because they could not get along with others? Explain your reasoning.

15. What fraction of the people represented in the chart were fired for reasons other than incompetence? Explain more than one way that you could find the answer to this question.

16. If the chart represents 5000 people, about how many were fired for dishonesty or lying? Explain.

Extensions

17. While Yolanda was at a carnival, she watched a game in which a paper cup was tossed. If the cup landed upright, the player won $5. It cost $1 to play the game. Yolanda watched the cup being tossed 50 times. The cup landed on its side 32 times, upside down 13 times, and upright 5 times.

   a. If Yolanda plays the game 10 times, about how many times can she expect to win? How many times can she expect to lose?

   b. Would you expect Yolanda to have more or less money at the end of 10 games than she had before? Why?

Mathematical Reflections

In this investigation, you conducted an experiment that involved tossing marshmallows. You also experimented with a coin-tossing game to determine whether it was fair. These questions will help you summarize what you have learned:

1. When you toss a large marshmallow, is it equally likely to land on an end as its side? What evidence can you use to help you answer this question?

2. How would you use the results of your work in Problem 2.1 to predict how many times a small marshmallow would land on its side if you tossed it 1000 times?

3. What does it mean for a two-person game of chance to be fair?

4. In a–f, give an example of an event that would have about the given chances of occurring.

   a. 0%   **b.** 10%
   c. 25% **d.** 50%
   e. 75% **f.** 100%

Think about your answers to these questions, discuss your ideas with other students and your teacher, and then write a summary of your findings in your journal.

Investigation 2: More Experiments with Chance

INVESTIGATION 3

Using Spinners to Predict Chances

School is out for the summer! Kalvin thinks he should be allowed to stay up until midnight every night since he doesn't have to get up for school in the morning. His father disagrees; he thinks Kalvin will have more energy for all the things he plans to do in the summer if he goes to bed earlier.

3.1 Bargaining for a Better Bedtime

Kalvin decided to make a spinner that he hopes his father will let him use to determine his bedtime each night. To encourage his father to go for his idea, Kalvin put three 10:00 and three 11:00 spaces on the spinner. However, he used the biggest space for 12:00, and he hopes the spinner will land on that space most often. Kalvin's spinner is shown on the next page.

> **Problem 3.1**
>
> Conduct an experiment to help you answer these questions.
>
> **A.** Kalvin prefers to go to bed at midnight, so he wants his spinner to land on 12:00 more often than anywhere else. Is it likely that this spinner will allow him to achieve this goal? Explain.
>
> **B.** Suppose Kalvin's father lets him use this spinner to determine his bedtime. What are Kalvin's chances of going to bed at 12:00? Explain how you determined your answer.
>
> To conduct your experiment, use Labsheet 3.1 and a bobby pin or paper clip to make a spinner like Kalvin's. Spin the spinner, and keep track of the data you collect. Continue spinning the spinner and recording data until you are confident about your answers to the questions above.

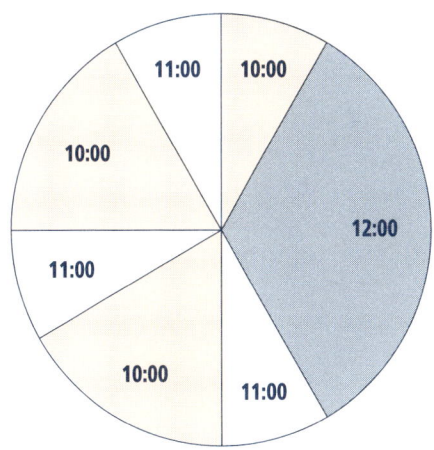

■ Problem 3.1 Follow-Up

1. After how many spins did you decide to stop spinning? Why? If you continued to spin the spinner, do you think your answers to Problem 3.1 would change? Why or why not?

2. **a.** How many times did you spin the spinner? How many times did the spinner land on 10:00? On 11:00? On 12:00?
   b. Based on your data, what fraction of the time will Kalvin go to bed at 10:00? At 11:00? At 12:00?
   c. Summer vacation is 90 days long. If Kalvin uses this spinner every night, how many nights do you think he will go to bed at 10:00? At 11:00? At 12:00? Explain your reasoning.

3. In a–c, use your angle ruler or other ways of reasoning to analyze Kalvin's spinner. You can set your angle ruler on the spinner to measure the angle of each section.

   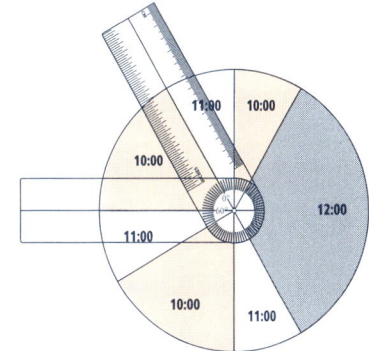

   a. What fraction of the area of the spinner is made up of 10:00 spaces? Of 11:00 spaces? Of 12:00 spaces?
   b. How do the fractions from part a compare with the fractions you found in part b of question 2?
   c. How do the fractions from part a compare with the fractions from the data your entire class collected for Problem 3.1?

Investigation 3: Using Spinners to Predict Chances

Applications • Connections • Extensions

As you work on these ACE questions, use your calculator whenever you need it.

Applications

1. In a–g, use the spinner on Labsheet 3.ACE.

   a. Use a paper clip or bobby pin to spin the spinner 30 times. What fraction of your spins landed on a space with hearts? With dots? With stripes?

   b. Use your angle ruler or another method to analyze the spinner. What fraction of the spinner is covered with hearts? With dots? With stripes? Explain how you found each fraction.

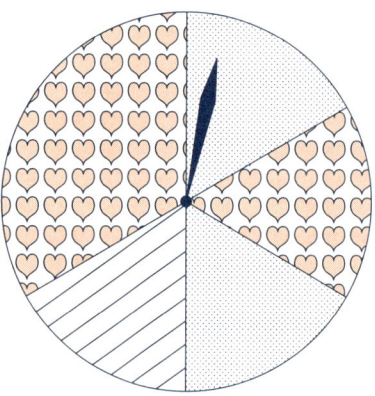

   c. Compare your answers to parts a and b. Would you expect these answers to be the same? Why or why not?

   d. If you were to spin the spinner 300 times instead of 30 times, do you think your answers would become closer to or further from the fractions you found in part b? Explain your reasoning.

   e. When you spin the spinner, are the three possible events—landing on a space with hearts, landing on a space with dots, and landing on a space with stripes—equally likely? Explain.

   f. Suppose you use the spinner to play a game with a friend. Your friend scores a point every time the spinner lands on a space with hearts. What spaces should you score on to make the game fair? Explain your reasoning.

   g. Suppose you use this spinner to play a three-person game. Player A scores if the spinner lands on stripes. Player B scores if the spinner lands on hearts. Player C scores if the spinner lands on dots. How could you allocate points so the game would be fair?

2. Mollie is designing a game for a class project. She made the three spinners shown here and experimented with them to see which one she liked best for her game. She spun each spinner 20 times and wrote down her results, but she forgot to record which spinner gave which set of data. Which spinner most likely gave each data set? Explain your answer.

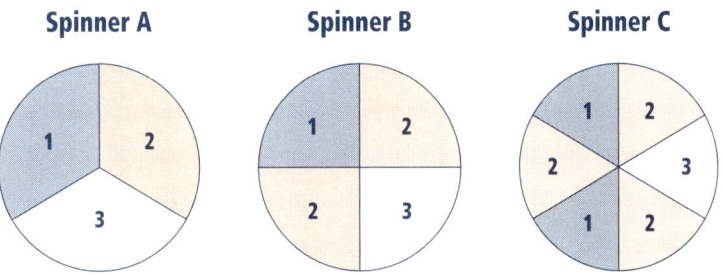

First data set
1 2 3 2 1 1 2 1 2 2 2 3 2 1 2 2 2 3 2 2

Second data set
2 3 1 1 3 3 3 1 1 2 3 2 2 2 1 1 1 3 3 3

Third data set
1 2 3 3 1 2 2 2 3 2 1 2 2 2 3 2 2 3 2 1

3. Three people play a game on each of the spinners in question 2. Player 1 scores a point if the spinner lands on an area marked 1, player 2 scores a point if the spinner lands on an area marked 2, and player 3 scores a point if the spinner lands on an area marked 3.

   a. On which spinner or spinners would the game be a fair game of chance? Why?

   b. Choose a spinner that you think would not make a fair game of chance with these rules. Then, change the scoring rules to make the game fair by assigning different points for landing on the different numbers. Explain why your point system works.

Investigation 3: Using Spinners to Predict Chances

4. a. Create a spinner and a set of rules for a two-person game that would be a fair game.

b. Create a spinner and a set of rules for a two-person game that would not be fair. Explain how you could change the rules to make the game fair.

Connections

In 5–9, use the data below to answer the question. If there is not enough information to answer a question, explain what additional information you would need.

> - In 1988, 47,093 people were killed in car crashes and 3486 people were killed in motorcycle crashes in the United States.
> - In the U.S., 40% of all deaths of people between the ages of 15 and 19 result from motor-vehicle crashes. Alcohol is involved in about half of these crashes.
> - Males outnumber females as fatal-crash victims by an average of 2 to 1.
> - 55% of motorcycle deaths occurred on weekends.
> - In 1988, the car with the lowest death rate was the Volvo 740/760 four-door, while the car with the highest death rate was the Chevrolet Corvette.
>
> Source: Michael D. Shook and Robert L. Shook, *The Book of Odds* (New York: Penguin Books, 1991), p. 90.

5. Which is safer to drive, a car or a motorcycle?

6. What percent of all deaths of 15-year-olds to 19-year-olds result from alcohol-related motor-vehicle crashes?

7. Is a particular motorcycle rider more likely to be in a fatal crash during the week or during the weekend?

8. Are males worse drivers than females?

9. Your family is trying to decide which used car to buy. Are you less likely to have an accident if you buy a Volvo 740 or Volvo 760 than if you buy a Chevrolet Corvette?

Extensions

10. Design a spinner with five spaces so that the chances of landing in each space are equally likely. Give the number of degrees in the central angle of each space.

11. Design a spinner with five spaces so that the chances of landing in each space are not equally likely. Give the number of degrees in the central angle of each space.

12. Design a spinner with five spaces so that the chances of landing in one space are twice the chances of landing in each of the other four spaces. Give the number of degrees in the central angle of each space.

Mathematical Reflections

In this investigation, you experimented with spinners. When you spin a spinner, you cannot know in advance which section it will land on, but you can conduct an experiment to gather data that will help you to predict what will happen over many trials. These questions will help you summarize what you have learned:

1 Suppose that out of 400 spins, a spinner lands 306 times on region A and 94 times on region B. What can you say about the spinner? What might the spinner look like? How confident are you in your answer? Explain.

2 Suppose that out of 20 spins, a spinner lands 13 times on region A and 7 times on region B. What can you say about the spinner? What might the spinner look like? How confident are you in your answer? Explain.

3 Describe how you could construct a spinner with four equally likely outcomes.

4 Look back at Kalvin's bedtime spinner for Problem 3.1. Is it possible that the spinner will land on 12:00 each night for a month? Is it likely?

Think about your answers to these questions, discuss your ideas with other students and your teacher, and then write a summary of your findings in your journal.

INVESTIGATION 4

Theoretical Probabilities

In the last three investigations, you worked with problems involving the chances that a particular event would occur. Another word for chance is *probability*. So far, you have determined probabilities by doing experiments and collecting data. For example, you flipped a coin many times and found that the probability of getting a head is $\frac{1}{2}$. You also discovered that the more trials that were done, the better the probabilities that you found could predict future outcomes.

The results of the coin-flipping experiment probably did not surprise you. You already knew that when a coin is flipped there are two possible outcomes—heads and tails—and that each outcome is equally likely. In fact, you could have found the probability of getting a head by *analyzing* the possible outcomes instead of by *experimenting*. Since there are two equally likely outcomes, and one of these outcomes is a head, the probability of getting a head is 1 out of 2, or $\frac{1}{2}$.

In this investigation, you will look at some other situations in which you can find the probabilities both by experimenting and by analyzing the possible outcomes.

4.1 Predicting to Win

In the last 5 minutes of the Gee Whiz Everyone Wins! television game show, all the members of the studio audience are called to the stage to select a block randomly from a bucket containing an unknown number of red, yellow, and blue blocks. Before drawing, each contestant is asked to predict the color of the block he or she will draw. If the guess is correct, the contestant wins a prize. After each draw, the block is put back into the bucket.

Think about this!

Suppose you are a member of the audience. Is there an advantage to being called to the stage first? Is there an advantage to being called last? Why?

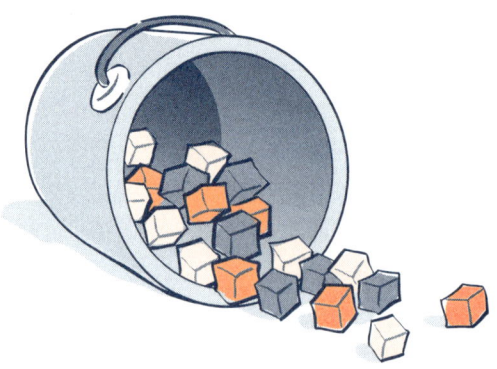

Problem 4.1

Play the block-guessing game with your class. Your teacher will act as the host of the game show, and you and your classmates will be the contestants. Keep a record of the number of times each color is drawn. Play the game until you think you can predict with certainty the chances of each color being drawn.

A. In your class experiment, how many blue blocks were drawn? Red blocks? Yellow blocks? What was the total number of blocks drawn?

B. The probability of drawing a red block can be written as P(red). Find all three probabilities based on the data you collected in your experiment.

P(red) =   P(yellow) =   P(blue) =

Now, your teacher will dump out the blocks so you can see them.

C. How many of the blocks are red? Yellow? Blue? How many blocks are there altogether?

D. Find the fraction of the total blocks that are red, the fraction that are yellow, and the fraction that are blue.

■ **Problem 4.1 Follow-Up**

The probabilities you computed in part B are called **experimental probabilities** because you found them by experimenting. The fractions you found in part D are called **theoretical probabilities.** You find theoretical probabilities by analyzing the possible outcomes rather than by experimenting.

If all the outcomes of an action are equally likely, then the theoretical probability of an event is computed with this formula:

$$\frac{\text{number of favorable outcomes}}{\text{number of possible outcomes}}$$

where *favorable outcomes* are the outcomes in which you are interested.

For example, if you want to find the probability of drawing a red block, a red block is a favorable outcome. If a bucket has a total of six blocks, and two of the blocks are red, the theoretical probability of drawing a red block is $\frac{2}{6}$.

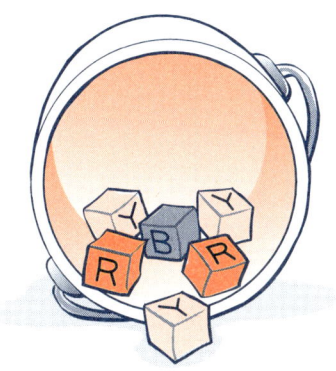

6 possible outcomes (blocks)
2 favorable outcomes (red blocks)

Theoretical probability of drawing a red block $= \frac{2}{6}$

1. Compare the *experimental probabilities* you found in part B to the *theoretical probabilities* you found in part D. Are the experimental and theoretical probabilities for each color of block close to each other? Do you think they should be close? Why or why not?
2. **a.** When you drew a block from the bucket, did each *block* have an equally likely chance of being chosen? Explain.
   b. When you drew a block from the bucket, did each *color* have an equally likely chance of being chosen? Explain.
3. Look back at the "Think about this!" on page 30. Is there an advantage to being the first person to draw from the bucket? To being the last person to draw?
4. In the Gee Whiz Everyone Wins! game show, contestants select a block randomly from the bucket. What do you think *random* means?

Investigation 4: Theoretical Probabilities

4.2 Drawing More Blocks

Your teacher put eight blocks in a bucket. All the blocks are the same size. Three are yellow, four are red, and one is blue.

Problem 4.2

A. When you draw a block from the bucket, are the chances equally likely that it will be yellow, red, or blue? Explain your answer.

B. What is the total number of blocks? How many blocks of each color are there?

C. What is the *theoretical probability* of drawing a blue block? A yellow block? A red block? Explain how you found each answer.

Now, as a class or in groups, take turns drawing a block from the bucket. After each draw, return the block to the bucket. Keep a record of the blocks that are drawn. If you work in a group, take turns drawing blocks until you have 40 trials.

D. Based on your data, what is the *experimental probability* of drawing a blue block? A yellow block? A red block?

E. Compare the theoretical probabilities you found in part C to the experimental probabilities you found in part D. Are the probabilities for each color close? Are they the same? If not, why not?

■ **Problem 4.2 Follow-Up**

Suppose you and your classmates each took three turns drawing a block from the bucket, replacing the block each time, and then used the large amount of data you collected to find new experimental probabilities for drawing each color. You found the theoretical probability of drawing each color in part C. Do you think these new experimental probabilities would be closer to the theoretical probabilities than the experimental probabilities you found in part D were? Explain your reasoning.

4.3 Winning the Bonus Prize

All the winners from the Gee Whiz Everyone Wins! game show get an opportunity to compete for a bonus prize. Each winner draws one block from each of two bags, both of which contain one red, one yellow, and one blue block. The contestant must predict which color she or he will draw from each of the two bags. If the prediction is correct, the contestant wins a $10,000 bonus prize!

Bag 1   Bag 2

Problem 4.3

What are a contestant's chances of winning?

Conduct an experiment to help you answer this question. Keep track of the pairs of colors that are drawn, and make sure you collect enough data to give you good estimates of the probability of drawing each pair. Remember, contestants must guess the color of the block they will pick from each bag. That means you will have to count (a blue from bag 1, a red from bag 2) as a different pair from (a red from bag 1, a blue from bag 2).

A. Based on your experiment, what are a contestant's chances of winning?

B. List all the possible pairs that can be drawn from the bags. Are each of these pairs equally likely? Explain your answer.

C. What is the theoretical probability of each pair being drawn? Explain your answer.

D. How do the theoretical probabilities compare with your experimental probabilities? Explain any differences.

Investigation 4: Theoretical Probabilities

Problem 4.3 Follow-Up

Suppose you are a contestant on the show, and you have already won a mountain bike, a fantastic portable CD player, a vacation to Hawaii, and a one-year supply of Glimmer toothpaste. You have just played the bonus round and lost, but the host makes the following offer: you can draw from the two bags again, but this time you do not need to predict the color. If the two colors match, you will win $5000. If the two colors do not match, you must return all the prizes you have won. Would you accept this offer? Explain why or why not.

Applications • Connections • Extensions

As you work on these ACE questions, use your calculator whenever you need it.

Applications

1. A bucket contains one green block, one red block, and two yellow blocks.

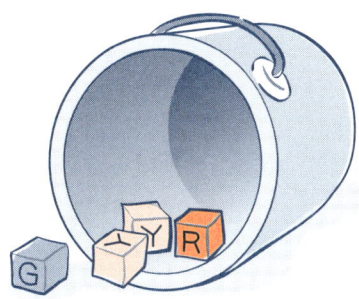

   a. Find the theoretical probability of choosing each color.

   P(green) = _____   P(yellow) = _____   P(red) = _____

   b. Find the sum of the probabilities in part a.

   c. What is the probability of *not* drawing a red block? Explain how you found your answer.

   d. What do you get when you add the probability of *getting* a red to the probability of *not getting* a red?

   e. What happens to the probability of drawing a red block if the number of blocks of each color is doubled?

   f. What happens to the probability of drawing a red block if two more blocks of each color are added to the original bucket?

   g. How many blocks of which colors would you have to add to the original bucket to make the probability of drawing a red block $\frac{1}{2}$?

Investigation 4: Theoretical Probabilities

2. A bag contains exactly three blocks, all blue.

   a. What is the probability of drawing a blue block?

   b. What is the probability of *not* drawing a blue block?

   c. What is the probability of drawing a yellow block?

3. A bubble-gum machine contains 25 gum balls. There are 12 green, 6 purple, 2 orange, and 5 yellow gum balls.

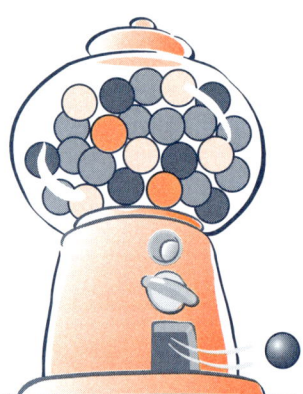

   a. Find the theoretical probability of getting each color.

   P(green) = _____   P(purple) = _____

   P(orange) = _____   P(yellow) = _____

   b. What is the sum of the probabilities for all the possible colors?

   P(green) + P(purple) + P(orange) + P(yellow) = _____

   c. Write each of the probabilities in part a as a percent.

   P(green) = _____   P(purple) = _____

   P(orange) = _____   P(yellow) = _____

   d. What is the sum of all the probabilities as a percent?

   e. What do you think the sum of the probabilities for all possible outcomes must be for any situation?

4. a. What do you think the word *probability* means?

   b. Describe some situations in which probability is important.

5. a. If two people do an experiment to estimate the probability of a particular event occurring, will they get the same result? Explain why or why not.

b. If two people analyze a situation to find the theoretical probability of an event occurring, and each person does a correct analysis, will they get the same result? Explain why or why not.

c. If one person uses an experiment to estimate the probability of an event occurring and another person analyzes the situation to find the theoretical probability of the event occurring, will they get the same result? Explain why or why not.

Connections

6. A bag contains several marbles. Some are red, some are white, and some are blue. Carlos counted the marbles and found that the theoretical probability of drawing a red marble is $\frac{1}{6}$ and the theoretical probability of drawing a white marble is $\frac{1}{3}$.

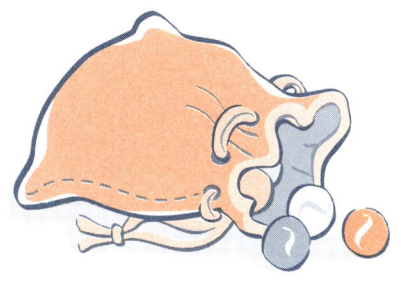

a. What is the smallest number of marbles that could be in the bag?

b. Could the bag contain 48 marbles? If so, how many of each color must it contain?

c. If the bag contains 4 red marbles and 8 white marbles, how many blue marbles must it contain?

d. How can you tell what the probability of a drawing a blue marble is?

7. Katherine's class made this line plot of the first letters in the first names of all the students in her class.

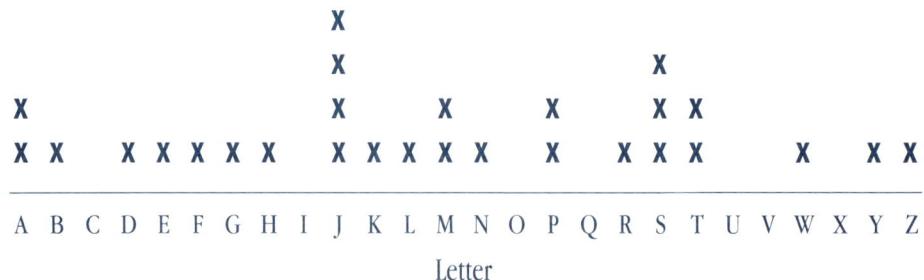

- **a.** If you randomly select a student from Katherine's class, what is the probability you will choose someone whose first name begins with J?

- **b.** If you randomly select a student from Katherine's class, what is the probability you will choose someone whose first name begins with a letter that occurs after F in the alphabet, but before T?

- **c.** If you randomly select a student from Katherine's class, what is the probability that you will choose Katherine?

- **d.** Suppose two more people joined the class, Melvin and Theo. Now if you randomly select a student from the class, what is the probability you will choose someone whose first name begins with J?

8. Suppose you were to spin this spinner and then roll this six-sided number cube.

a. Make an organized list of the possible outcomes of a spin of the spinner and a roll of the number cube. For example, the outcome that is showing is this:

   Spinner   Number cube
      2   2

b. What is the probability you would get a 2 on both the number cube and the spinner? Explain your reasoning.

c. What is the probability you would get a *factor* of 2 on both the number cube and the spinner?

d. What is the probability you would get a *multiple* of 2 on both the number cube and the spinner?

Extensions

9. The cook in the Casimer Middle School cafeteria is in a bad mood! When Jonalyn went through the lunch line, she tried to tell the cook what she wanted, but the cook just mumbled, "Appreciate what you get!" Jonalyn thinks some of the things on the menu are really gross. Her favorite lunch is a grilled cheese sandwich, carrots, and a chocolate chip cookie.

Investigation 4: Theoretical Probabilities

Lunch at Casimer consists of one sandwich, one vegetable, and one cookie. The cook has an equal number of each kind of sandwich, vegetable, and cookie. She is not paying any attention to how she puts the lunches together.

a. How many different lunches are possible? Explain your answer.

b. What is the probability that Jonalyn will get her favorite lunch? Explain your reasoning.

c. What is the probability that Jonalyn will get at least *one* of her favorite things? Explain your reasoning.

10. Make up a bag containing 12 objects—such as blocks or marbles—of the same size and shape. Use three or four different colors.

a. Describe the contents of your bag.

b. Determine the *theoretical probability* of drawing each color by analyzing the bag's contents.

c. Conduct an experiment to determine the *experimental probability* of drawing each color. Carefully describe how you did your experiment and recorded your results.

d. How do the two types of probability you found compare?

Mathematical Reflections

In this investigation, you studied a new way to find probabilities. You now have two ways to get information about the chances, or probability, that something will occur. You can design an experiment and collect data (to find experimental probabilities), or you can think about a situation, analyzing the outcomes carefully to see exactly what might happen (to find theoretical probabilities). These questions will help you summarize what you have learned:

1. How can you find the experimental probability of an event? Why is this called an *experimental probability*?

2. How can you find the theoretical probability of an event? Why is this called a *theoretical probability*?

3. When you tossed coins to figure out which cereal Kalvin would have for breakfast, you found experimental probabilities by counting heads and tails. You might have noticed that the more trials you did, the closer your experimental probability came to the theoretical probability of $\frac{1}{2}$. Do you think that conducting more trials will always bring your experimental probability closer to the theoretical probability? Why or why not?

4. Think of some situations in which it would be easier to find theoretical probabilities than experimental probabilities. Explain your reasoning.

5. Think of some situations in which it would be easier to find experimental probabilities than theoretical probabilities. Explain your reasoning.

Think about your answers to these questions, discuss your ideas with other students and your teacher, and then write a summary of your findings in your journal.

Investigation 4: Theoretical Probabilities

INVESTIGATION 5

Analyzing Games of Chance

Have you ever figured out a strategy for winning a game? In this activity, you will play a two-team game called Roller Derby. As you play, think about strategies for winning and the probabilities associated with those strategies.

5.1 Playing Roller Derby

In a game of Roller Derby, two teams compete. Each team needs a game board with columns numbered 1 through 12, a pair of number cubes, and 12 markers (like pennies, buttons, or small blocks).

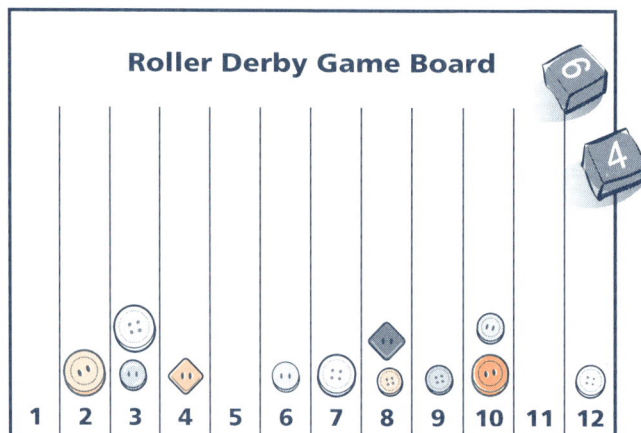

Roller Derby Rules

1. Each team places its 12 markers into the columns in any way it chooses.
2. Each team rolls a number cube. The team with the highest roll goes first.
3. Teams take turns rolling the two number cubes and removing a marker from the column with the same number as the total shown on the cubes. If the column is empty, the team does not get to remove a marker.
4. The first team to remove all the markers from its board wins.

Problem 5.1

What is a good strategy for placing your markers in the 12 columns on the game board?

Play the game at least twice before answering this question. As you play, keep a record of the strategies you use.

■ Problem 5.1 Follow-Up

1. **a.** Find a systematic way to list all the possible outcomes (number pairs) of rolling two number cubes and the sums for each of these outcomes. Analyze your list carefully before answering b–e.
   b. What sums are possible when you roll two cubes?
   c. Which sum or sums occur most often?
   d. How many ways can you get a sum of 6? A sum of 2?
   e. Are all the sums equally likely? Explain.
2. Now that you have analyzed the possible outcomes, do you have any new ideas for a strategy for winning Roller Derby? Explain. If time allows, play the game again using your new strategy.

Applications • Connections • Extensions

As you work on these ACE questions, use your calculator whenever you need it.

Applications

1. Eleanor is playing Roller Derby with Carlos. Eleanor placed all of her markers in column 1, and Carlos placed all of his markers in column 12. What is the probability that Eleanor will win? What is the probability that Carlos will win? Explain your reasoning.

2. When you play the game of Monopoly®, you sometimes end up in "jail." One way to get out of jail is to roll a double (two cubes that match). What is the probability of getting out of jail on your turn by rolling a double? Use your list of possible outcomes of rolling two number cubes to help you answer this question. Explain your reasoning.

Connections

In 3–9, use your list of possible outcomes of rolling two number cubes to help you answer the question.

3. When two number cubes are rolled, what is the probability that their sum will be 3?

4. When two number cubes are rolled, what is the probability that their sum will be greater than 9?

5. When two number cubes are rolled, what is the probability that their sum will be a multiple of 4?

6. When two number cubes are rolled, what is the probability that their sum will be a common multiple of 2 and 3?

7. When two number cubes are rolled, what is the probability that their sum will be a prime number? Explain.

8. Which has a greater probability of being rolled on a pair of number cubes—a sum that is a factor of 6 or a sum that is a multiple of 6? Explain.

9. Humberto and Kate are playing a game called Odds and Evens. To play the game, they roll two number cubes. If the sum is odd, Humberto scores a point. If the sum is even, Kate scores a point. Is this a fair game of chance? Why or why not?

10. Suppose that Humberto and Kate play a game called Evens and Odds. (This game is similar to the game in question 9, except it involves *products* instead of *sums*.) To play the game, they roll two number cubes. If the product is odd, Kate scores a point. If the product is even, Humberto scores a point.

    a. Make an organized table of the possible products of two number cubes.

    b. What is the probability that Kate will win? What is the probability that Humberto will win? Explain your reasoning.

    c. Is this a fair game? If it is fair, explain why. If it is not fair, tell how you could change the points scored by each player so that it would be fair?

    d. What is the probability that the product rolled will be a prime number?

    e. What is the probability that the product rolled will be a factor of 30?

    f. What is the probability that the product rolled will be greater than 18?

Investigation 5: Analyzing Games of Chance

11. The cooks at Kyla's school made the spinners shown below to help them determine the lunch menu. They let the students take turns spinning to determine the daily menu. In a–c, decide which spinner you would choose, and explain your reasoning.

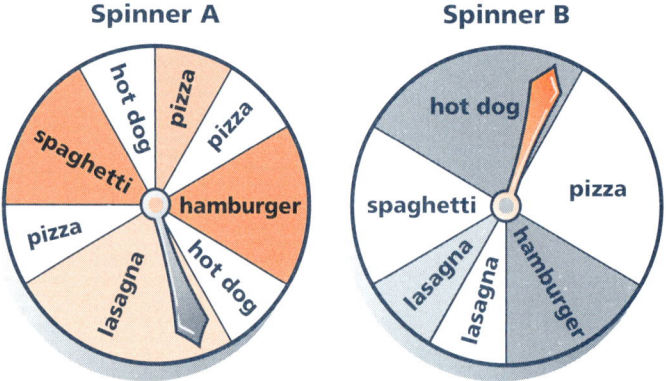

Spinner A **Spinner B**

a. Your favorite lunch is pizza.

b. Your favorite lunch is lasagna.

c. Your favorite lunch is hot dogs.

12. Abigail and Christopher are playing a game with two coins. To play the game, they each flip a coin at the same time. If the two coins match, Christopher gets a point; if they do not match, Abigail gets a point. Is this a fair game of chance? Explain your reasoning.

13. Alex and Fumi are playing a game with three coins. To play the game, they flip all three coins at the same time. If the three coins match, Fumi gets a point. If they do not all match, Alex gets a point. Is this a fair game of chance? Explain your reasoning.

46 How Likely Is It?

Extensions

14. Make up three probability questions that can be answered by looking at your list of possible outcomes of rolling two number cubes. Then answer your own questions.

In 15–18, suppose Selina has just rolled three number cubes.

15. What is the probability that all three cubes match? Explain your reasoning.

16. What is the probability that the sum of the cubes is less than 5? Explain your reasoning.

17. What is the probability that the sum of the cubes is more than 2? Explain your reasoning.

18. What is the probability that the product of the cubes is prime? Explain your reasoning.

Mathematical Reflections

In this investigation, you played a game of chance that involved rolling a pair of number cubes and computing the sum of the cubes. These questions will help you summarize what you have learned:

1. What are the possible outcomes when you roll one number cube? Is each of these outcomes equally likely?

2. When you roll a pair of number cubes, how many different pairs of numbers can occur? Is each pair equally likely?

3. In the Roller Derby game, you added the numbers on the faces of two number cubes. How many different sums were possible? Were they all equally likely? Explain.

4. What is the sum of the probabilities of all the outcomes you can get when you roll two number cubes? Explain your answer.

Think about your answers to these questions, discuss your ideas with other students and your teacher, and then write a summary of your findings in your journal.

INVESTIGATION 6

More About Games of Chance

Have you ever tried to win a contest? Stores and restaurants often have contests to attract customers. Knowing something about probability can often help you figure out your chances of winning these contests.

6.1 Scratching Spots

Tawanda's Toys is having a contest! Any customer who spends at least $10 receives a scratch-off game card. Each card has five gold spots that reveal the names of video games when they are scratched. Exactly two spots match on each card. A customer may scratch off only two spots on a card; if the spots match, the customer wins the video game under those spots.

Problem 6.1

If you play this game once, what is your probability of winning? To answer this question, do the following two things:

A. Create a way to simulate Tawanda's contest, and find the experimental probability of winning.

B. Analyze the different ways you can scratch off two spots, and find the theoretical probability of winning a prize with one game card.

■ Problem 6.1 Follow-Up

1. **a.** If you play Tawanda's scratch-off game 100 times, how many video games would you expect to win?

   b. How much money would you have to spend to play the game 100 times?

2. Tawanda wants to be sure she will not lose money on her contest. The video games she gives as prizes cost her about $15 each. Will Tawanda lose money on this contest? Why or why not?

3. Suppose you play Tawanda's game 20 times and never win. Would you conclude that the game is unfair? For example, would you think that there were not two matching spots on every card? Why or why not?

Applications • Connections • Extensions

As you work on these ACE questions, use your calculator whenever you need it.

Applications

1. Tawanda thinks there should be fewer winners in her contest. She has decided to order new cards with six spots. Two of the spots on each card match. What is the probability that a person who plays the game once will win a prize?

2. The Kalikak High School Science Club is hosting a carnival to raise money for a trip to the national science fair in San Diego, California. They will have a game called Making Purple at the carnival. The game involves the two spinners below. A player spins spinner A and spinner B. If the player gets red on spinner A and blue on spinner B, the player wins, because red and blue together make purple.

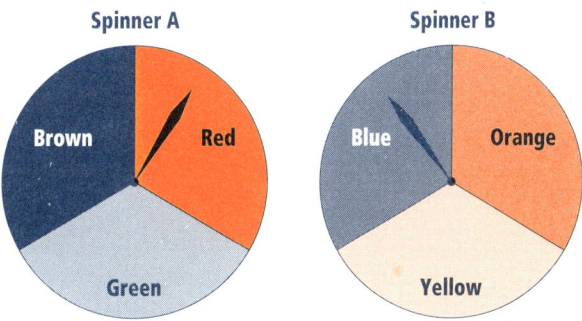

   a. List the outcomes that are possible when both spinners are spun. Are these outcomes equally likely? Explain your reasoning.

   b. What is the theoretical probability that a player will "make purple"? Explain.

   c. The science club will charge $1.00 per spin. A player who makes purple wins $5.00. If 100 people play this game, how many people would you expect to win? Explain your reasoning.

   d. If 100 people play, how much money would you expect the science club to make?

Investigation 6: More About Games of Chance

Connections

3. The Federal Trade Commission is the part of the U.S. government that makes rules for businesses that buy and sell things. The Federal Trade Commission Act states that an advertisement may be found unlawful if it could deceive someone.

The FTC doesn't need to prove that anyone was actually deceived by an advertisement to decide that it is deceptive and unlawful. To decide whether an ad is deceptive, the FTC considers the "general impression" it makes on a "reasonable person." So even if every statement in an ad is true, the ad is deceptive if it gives an overall false impression. (For example, companies cannot show cows in margarine commercials, because it gives the false impression that margarine is a dairy product.)

a. Suppose Tawanda placed this advertisement in the newspaper.

According to the Federal Trade Commission Act, do you think it is legal for Tawanda to say that "every card is a winner"? Explain your answer.

b. Design a better advertisement for Tawanda that will make people excited about the contest but will not lead some to think they will win every time they play.

4. A sugarless gum company used to have an advertisement that stated:

> *Nine out of ten dentists surveyed recommend sugarless gum for their patients who chew gum.*

Do you think this statement means that 90% of dentists think their patients should chew sugarless gum? Explain your reasoning.

5. Suppose you are the coach of the U.S. all-star baseball team. You need to pick someone to pinch hit for the pitcher. You look over the records of your players and narrow your choices to these three:

Player	At bats	Hits
George Brett	9789	3005
Kirby Puckett	5645	1812
Wade Boggs	6213	2098

Source: Mike Meserole (ed.) *1993 Sports Almanac* (Boston: Houghton Mifflin), p. 109.

a. What percent of Brett's at bats were hits? Puckett's? Boggs'?

b. Which player has the greatest chance of getting a hit on his next turn at bat? Explain your reasoning.

6. Willie Mae has flown over a million miles as an airline passenger without ever being in an accident. Kobie has never flown in an airplane. Both are planning to take a trip in an airplane.

   a. Who do you think is more likely to be in an airplane accident? Why?

   b. Does your answer to part a make sense if Willie Mae and Kobie get on the same airplane? Explain.

7. A-1 Trucks used this graph to show that their trucks last longer than other companies' trucks. A-1 Trucks is company A on the graph.

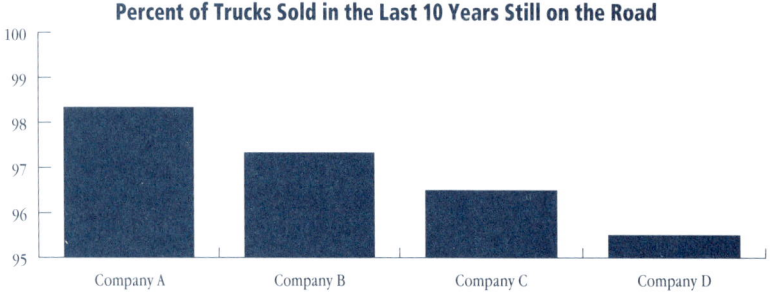

   a. The bar for company A is about six times as tall as the bar for company D. Does this mean that the chances of one of company A's trucks lasting 10 years are about six times as great as the chances of one of company B's trucks lasting 10 years? Explain your reasoning.

   b. If you wanted to buy a truck, would this graph convince you to buy a truck from company A? Why or why not?

Extensions

8. Refer to the discussion of the Federal Trade Commission Act in question 3 above. Find an advertisement that might be deceptive, and bring it in to discuss with your class. You might consider contacting the company, telling them why you think the ad might be deceptive, and asking for proof of their claims. The company is required by law to respond.

In 9–11, imagine that you help businesses by designing promotional contests. Design a contest for each company. Each contest should help the company attract customers, but not make the company lose money. For each contest, explain the rules, including any requirements for entering the contest, and design an advertisement for the contest.

9. The Fashion Gallery is a small clothing store. Its manager would like you to design a contest in which 1 of every 30 players wins a prize.

10. Supermart Superstores is a chain of supermarkets with over 100 locations. The director of operations would like to have a contest with a $100,000 grand prize!

11. Ally's AutoWorld sells new and used cars. Ally would like to have a contest with lots of winners and fairly big prizes. She would like about one out of every ten players to win a $500 prize.

Mathematical Reflections

In this investigation, you examined what you might *expect* to gain or lose when you play a game of chance. These questions will help you summarize what you have learned:

1 How can you find the number of times you would expect to win if you play Tawanda's game on a five-spot card 1000 times?

2 How can you find the number of times you would expect to win if you play Tawanda's game on a six-spot card 1000 times?

3 Suppose that your probability of winning a game at the school fair is $\frac{1}{6}$. It costs 10¢ to play the game, and the prize for winning is 50¢. Describe how you could decide whether this is a fair game of chance. (A *fair game of chance* is one in which you would expect to break even in the long run.)

Think about your answers to these questions, discuss your ideas with other students and your teacher, and then write a summary of your findings in your journal.

INVESTIGATION 7

Probability and Genetics

Have you ever wondered why your eyes and your hair are the color they are? Scientists who study traits such as eye and hair color are called *geneticists*. Geneticists use probabilities to predict the occurrence of certain traits in children based on the traits of their parents, grandparents, and other relatives.

One interesting genetic trait is the ability to curl the tongue into a U shape. In this investigation you will explore the question, What are the chances that someone can curl his or her tongue?

7.1 Curling Your Tongue

One day Kalvin was teasing his little sister Kyla, and he stuck his tongue out at her. She noticed that his tongue was curled into a U shape. Kyla said, "That's weird, Kalvin—your tongue looks goofy!"

Kalvin looked in the mirror and noticed that he *could* curl his tongue. He wondered how many other people can curl their tongues.

> **Problem 7.1**
>
> What fraction of students in your class can curl their tongues?
>
> With your class, conduct a survey of the students in the class to investigate tongue curling and to answer this question.

Problem 7.1 Follow-Up

What is the probability that a student you choose randomly from the hallway of your school will be able to curl his or her tongue?

7.2 Tracing Traits

Surveys are often used to gather information about a group of people, or a *population*. For example, if scientists want to find out the percent of people in a population that have a certain disease, they might conduct a survey of a large number of people. Sometimes scientists are interested in the probability that a *specific person* has a certain trait or will have a particular disease. In these situations, geneticists study the traits of the person's parents, grandparents, and other family members.

Have you every heard of *genes?* (We don't mean the kind you wear!) Your parents gave you a unique set of genes that determines many of your traits, such as your eye color, whether you are color blind, whether you will be bald someday, and whether you can curl your tongue.

Did you know?

Psychologists have long been interested in investigating how great a part genes play in determining human intelligence. One way of learning more about this topic is by studying identical twins who have been separated at a young age and raised in very different kinds of homes. Studies have shown that such twins showed remarkable similarities in intelligence. One study showed that twins raised in different home environments had IQ scores almost as close as those of identical twins raised in the same home. Since identical twins have the exact same set of genes, these results support the idea that genes play a key role in human intelligence.

Even more surprising was learning that these twins also were very similar in physical appearance, dress, mannerisms, preferences, attitudes, and even personality. Many of the twins had similar hairstyles, moved their hands in similar ways, or had the same attitude toward their jobs.

58 How Likely Is It?

Geneticists use the word *allele* to mean a special form of a gene. For example, you have two alleles that determine whether or not you can curl your tongue. Each of your parents also has two alleles for tongue curling. You received one of your alleles from your mother and one from your father. Each of your mother's two alleles had an equal chance of being passed on to you, and each of your father's two alleles had an equal chance of being passed on to you.

Let's let a capital T stand for the allele for tongue curling, and a small t stand for the allele for non-tongue curling. If a person receives a T allele from each parent, his tongue-curling alleles will be TT, and he will be able to curl his tongue. If a person receives a t allele from each parent, his tongue-curling alleles will be tt, and he won't be able to curl his tongue. What if a person receives one T allele and one t allele? Nature has figured out a way to break this tie. In the case of tongue curling, the T allele is *dominant,* and the t allele is *recessive.* This means that if a person has a Tt allele combination, the T allele dominates, and the person has the tongue-curling trait.

If your tongue-curling alleles are TT or Tt, you can curl your tongue. If your tongue-curling alleles are tt, you won't be able to curl your tongue—no matter how hard you try!

An Example: Bonnie and Ebert's Baby

Bonnie and Ebert are going to have a baby. Bonnie's tongue-curling alleles are Tt, and Ebert's tongue-curling alleles are tt. You can determine the probability that their baby will be able to curl his or her tongue. Here is a diagram of the allele possibilities for the baby:

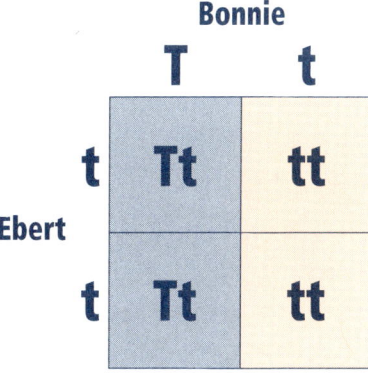

Ebert's alleles are shown at the left side, and Bonnie's alleles are shown at the top. The Tt in the upper-left square is the combination of Bonnie's T allele and Ebert's t allele.

You can see that there are four possible allele pairs (outcomes). Two of these pairs—Tt and Tt—result in the tongue-curling trait. So, the probability that Bonnie and Ebert's baby will have the tongue-curling trait is $\frac{2}{4}$ or $\frac{1}{2}$.

The probability of a child being able to curl his or her tongue will not always be $\frac{1}{2}$. If the parents' alleles are different from Ebert and Bonnie's, the probability will be different.

Problem 7.2

Kalvin's mother is pregnant with her third child. Kalvin figured out from studying his family for several generations that his mother and father both have the tongue-curling alleles Tt. Based on what you know about his parent's alleles, what is the probability that Kalvin's new sibling will be able to curl his or her tongue?

■ **Problem 7.2 Follow-Up**
1. Neither of Eileen's parents can curl their tongues. What is the probability that Eileen can curl her tongue?
2. Suppose that Geoff's tongue-curling alleles are TT and Mali's tongue-curling alleles are Tt. What is the probability that their child will be able to curl his or her tongue?
3. Marc can curl his tongue, and he wonders whether his parents can. He asks his mother to try it, and she can't curl hers. Do you think Marc's father can curl his tongue? Why or why not?
4. If Rodney's mother and father can both curl their tongues, can you conclude that Rodney can curl his tongue? Explain.

There are many other dominant traits that you can study in the way you have just studied tongue curling. For example, brown eyes are dominant over blue eyes, having a hairy head as an adult is dominant over having a bald head as an adult, and having a "hitchhiker's thumb" (also called a double-jointed thumb) is dominant over not having it.

Applications • Connections • Extensions

As you work on these ACE questions, use your calculator whenever you need it.

Applications

In 1–6, use the following information about the genetics of eye color to answer the question. The alleles for blue eyes and brown eyes work similarly to tongue-curling alleles. Let B stand for the brown-eyes allele, and let b stand for the blue-eyes allele. B is dominant, so a person with BB or Bb will have brown eyes, while a person with bb will have blue eyes. (You may have noticed that we have not talked about green eyes and other variations. These things can get pretty complicated—you might learn more about this in your high-school science classes.)

1. Suppose two blue-eyed people are expecting a baby. What is the probability that their child will have brown eyes? Explain.

2. Suppose a brown-eyed person with alleles BB and a blue-eyed person are expecting a baby. What is the probability that the baby will have brown eyes? Explain.

3. If Laura has brown eyes, could both of her parents have blue eyes? Why or why not?

4. If Katrina has blue eyes, could both of her parents have brown eyes? Why or why not?

5. Suppose Ken and Andrea both have brown eyes. They are wondering how many of their children will have brown eyes.

   a. Andrea's mother has brown eyes, and her father has blue eyes. What are Andrea's eye-color alleles? Explain.

   b. Ken's mother has blue eyes, and his father has brown eyes. What are Ken's eye-color alleles? Explain.

   c. What is the probability that Ken and Andrea's first child will have brown eyes?

   d. If their first child has brown eyes, what is the probability that their second child will also have brown eyes?

   e. Suppose Ken and Andrea have ten children. How many of their children would you expect to have brown eyes? Why?

Investigation 7: Probability and Genetics

6. Suppose you are a geneticist and you are trying to determine Dawn and Tomas's eye-color alleles. Here is the information you have:

- Dawn has blue eyes.
- Tomas has brown eyes.
- Their two daughters have brown eyes.
- Their son has blue eyes.

a. What are Dawn's eye-color alleles?

b. What are Tomas's eye-color alleles?

c. If they have another child, what is the probability that he or she will have blue eyes?

Connections

7. Write your own definition for the word *probability*. In your definition, show what you have learned about probability during this unit.

8. a. Write your own explanation about how experimental and theoretical probabilities are alike and different.

b. When you surveyed your classmates to find the probability that a student has the tongue-curling trait, were you finding an experimental or a theoretical probability? Explain.

c. When you found the probability that Kalvin's new sibling would have the tongue-curling trait, were you finding an experimental or theoretical probability? Explain.

Extensions

9. Pick one of the following two options:

   a. Investigate tongue curling in your family. Make a family tree that shows the tongue-curling alleles that you can figure out for each person. Trace back as many generations as you can. (If you'd like, you may do this for eye color instead.)

   b. Survey a large number of people to estimate the percent of people in the population who can curl their tongues. Represent the data in a graph. How do these data compare with your class's data?

Mathematical Reflections

In this investigation, you studied an example of a way that probabilities are used to predict a person's characteristics, such as eye color or tongue curling. These questions will help you summarize what you have learned:

1. How was probability used in your class's tongue-curling experiment?
2. How was probability used in your theoretical analysis of tongue curling?
3. If both parents of a child can curl their tongues, will the child be able to curl his or her tongue? Explain.

Think about your answers to these questions, discuss your ideas with other students and your teacher, and then write a summary of your findings in your journal.